Science Sight Word Readers™

Skin

by Casey Lasko

ISBN 978-0-545-24797-9

Photographs © 2010: cover: Getty Images/Flamingo Photography; back cover top: iStockphoto/vesilvio; back cover bottom: iStockphoto/TerrainScan; page 1: iStockphoto/Rosemarie Gearhart; page 2: iStockphoto/Jacek Chabraszewski; page 3 left: ShutterStock, Inc./Sebastian Kaulitzki; page 3 right: Photo Researchers, NY/Andrew Syred; page 4: iStockphoto/TerrainScan; page 5 left: Photo Researchers, NY/Phil Jude; page 5 right: iStockphoto/Pavel Losevsky; page 6 top: Masterfile; page 6 bottom: iStockphoto/Sebastian Kaulitzki; page 7: ShutterStock, Inc./Suzanne Tucker; page 8 top: iStockphoto/Kim Gunkel; page 8 bottom: iStockphoto/vesilvio; page 9 left: Getty Images/Ellen Denuto; page 9 right: iStockphoto/Valentin Casarsa; page 10: Corbis Images/ Richard Hutchings; page 11 left: iStockphoto/Sean Locke; page 11 right: ShutterStock, Inc./Tyler Olson; page 12 top: iStockphoto/ Steven Dern; page 12 bottom: iStockphoto/Ryan Lane; page 13: iStockphoto/Jani Bryson; page 14: iStockphoto/Amanda Mack; page 15: iStockphoto/Rosemarie Gearhart; page 16: iStockphoto/TerrainScan.

Photo research by Veroniqua Quinteros; Design by Holly Grundon

12 11 10 9 8 7 6 5 4 3 2 1   10 11 12 13 14 15/0
Printed in the U.S.A.   40
First printing, September 2010

SCHOLASTIC INC.

NEW YORK • TORONTO • LONDON • AUCKLAND
SYDNEY • MEXICO CITY • NEW DELHI • HONG KONG

It's **time** to **learn** about your **skin**!
Skin covers your whole body.

Under Your Skin

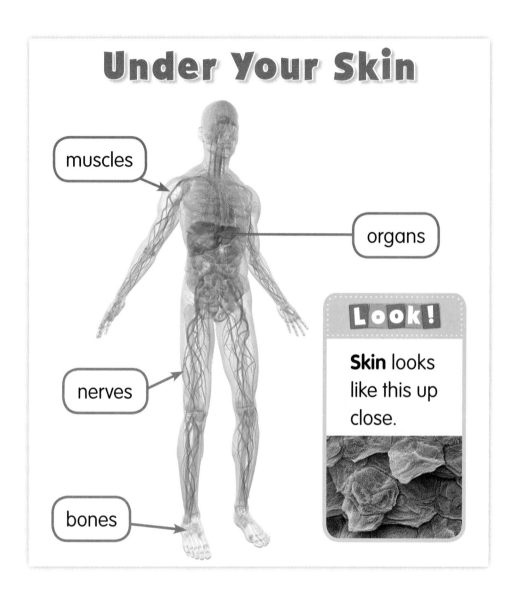

muscles

organs

nerves

bones

Look!

Skin looks like this up close.

Skin keeps your insides in.

Layers of Skin

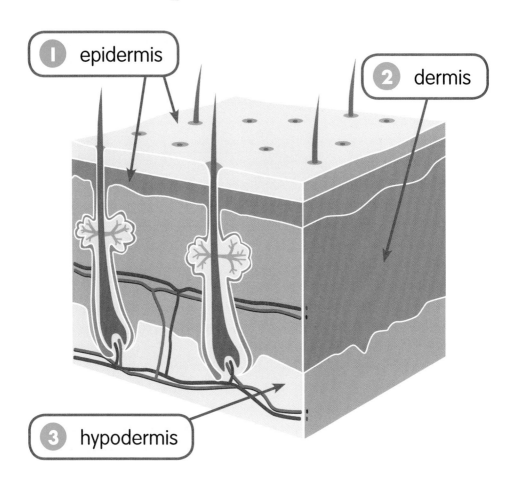

1 epidermis

2 dermis

3 hypodermis

It's time to **learn** about your **skin**!
Skin has layers.

Hair helps keep you warm.

Skin is covered with tiny hairs.

The nerve endings in your **skin** send signals to your brain. Ouch!

nerve

It's time to **learn** about your **skin**!
Skin can feel things.

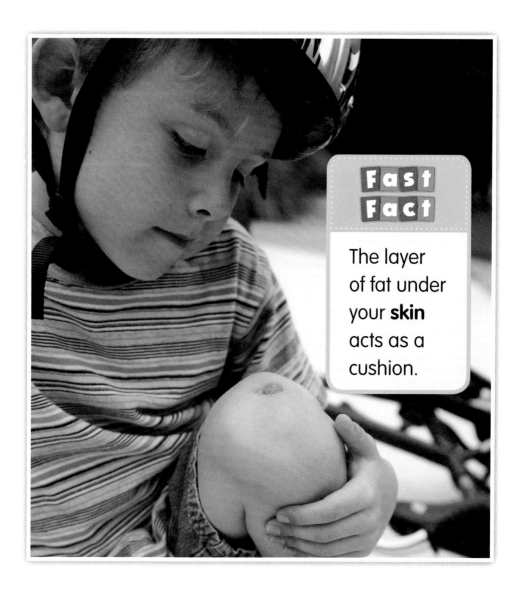

Skin helps to protect your insides from getting hurt.

It's **time** to **learn** about your **skin**!
Skin keeps out dirt and germs.

Skin keeps out water, too.

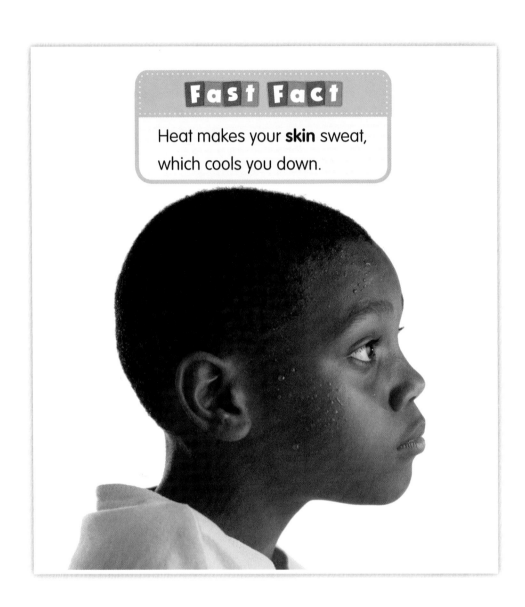

Fast Fact

Heat makes your **skin** sweat, which cools you down.

It's time to **learn** about your **skin**!
Skin changes when you're hot.

People get goose bumps on their **skin** when they are cold.

Skin changes when you are cold.

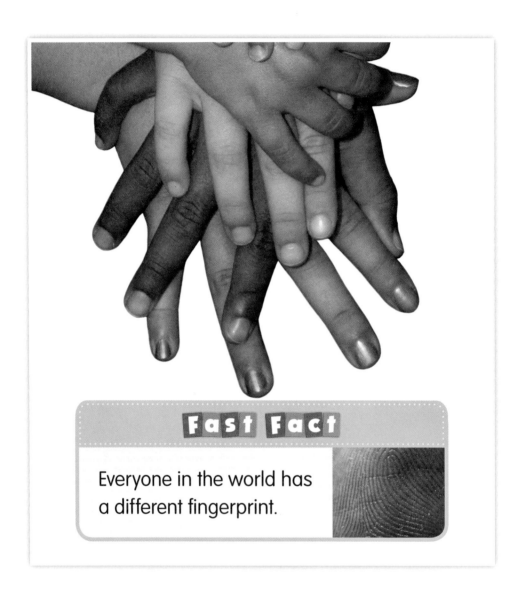

Fast Fact

Everyone in the world has a different fingerprint.

It's **time** to **learn** about your **skin**!
Skin comes in many colors.

Yippee! You **learn**ed all about your **skin**!

Sight Word Review

Point to each sight word. Then read it aloud.

learn

skin

it's

time

skin

it's

time

learn

Sight Word Fill-ins

Use one sight word from the box to finish each sentence.

it's	learn
skin	time

1 I like to _____ about the human body.

2 When your hands are dirty, it's _____ to wash them.

3 Your _____ keeps your insides in!

4 _____ important to keep your skin clean.

All About Skin

Ask a grown-up to read this with you.

Your body is covered in skin. It keeps your bones, muscles, and organs in. It keeps dirt and germs out.

Layers of Skin

1 epidermis
2 dermis
3 hypodermis

The skin on some parts of your body is thinner than paper. But one square inch of thin skin contains 19 million cells. Cells are the tiny building blocks that make up your body. Each day, about 40,000 dead skin cells flake off your body. But you are always making new skin cells. The skin you have today will be replaced by brand-new skin in about a month's time.

Your thin skin does a lot of different things. Your skin is packed with nerve endings. They warn you if something is hot or sharp.

Your skin is full of tiny tube-shaped sweat glands. When you get hot, sweat comes out of little holes in your skin called pores. Sweat cools you off.

Your skin is also covered with tiny hairs. If you get cold, the tiny hairs stand on end. The hairs pull on your skin to create goose bumps. Brrrr! You also get goose bumps if you're scared. Boo!